BIOLOGY TOPIC TEXT D: INHERITANCE AND EVOLUTION

HAROLD M. ICKES II, M.D.

Executive Editor: Rhett C. Smith, Ph.D.

Ideal Quill Publishing Group, LLC

idealquill.com

Printed in the United States of America

10 9 8 7 6 5 4 3 2 1

ISBN 978-0615834856

Biology Topic Text D: Inheritance and Evolution

by Harold M. Ickes II, M.D.

Table of Contents

Unit 1

Basic Inheritance

In this section we cover:

1) Genotype and phenotype
2) Principle of Segregation
3) Expression Patterns:
4) Partial Dominance Inheritance
5) Dominant/Recessive Inheritance
6) Sex-linked Inheritance
7) Maternal Inheritance

We are now in a position to start studying **inheritance**[1]. This is a rough road that will ultimately lead us to the central bastion of modern biology: evolution. We'll take this road one step at a time and learn the concepts carefully. To start, let's briefly review what we've covered in previous topic texts. DNA is a collection of 4 different kinds of base pairs. The order of these pairs is used as a code to hold information. Discrete stretches of base pairs that code for protein are called genes. Genes are surrounded by genetic "punctuation marks" that tell the cell's machinery when to start and stop reading. Messenger RNA is transcribed from the DNA in genes. This mRNA is then taken to ribosomes and translated using the genetic code to make protein, where the order of the codons in the messenger RNA determines the order of the amino acids in the protein. Proteins have many different roles in the cell; some help control chemical reactions

---

[1] Here, inheritance is defined as the passage of genetic material from parent to offspring at the time of fertilization, not the scraps we fight our siblings over when our parents pass away. I still can't believe my brother got the leg lamp...

(such as those in glycolysis and the citric acid cycle[2]), while others have structural roles, and so forth. Therefore, the order of the DNA in genes has a direct influence on the overall function of the protein, and through that function the DNA influences the performance of the cell and overall organism. This connection can be obvious, such as a gene that makes someone's eyes blue, or makes someone tall, or gives someone blond hair, and so forth, or the genes can have a subtle influence, such as making someone more or less susceptible to emotional problems, making someone vulnerable to some types of infections, etc. In these latter cases, it becomes difficult to say what effects are due to the genes and what are due to the environment. This is what the whole "Nature vs. Nurture" hubbub is all about.

Anyway, in genetics we say that an organism's genetic information (the order of its DNA) is its **genotype**, while the effect of that genetic information (eye color, hair color, etc.) is its **phenotype**. The phenotype that appears most commonly in the wild (i.e., natural settings) is called the **wildtype**. For example, suppose we monkeyed around with a mouse's DNA (put in a new genotype) so that we could get it to glow purple (a new phenotype). This new phenotype might certainly be quite fascinating, but it's far from the typical wildtype characteristics or normal dull grey fur.

Now that we've reviewed and covered these basic points, let's talk a little more on what homologous chromosomes are. We've mentioned them throughout the previous topic text, but their significance has not been particularly clear until now.

Homologous Chromosomes

For diploid organisms, every chromosome has a homologous partner. Now, if you'll recall, homologous chromosomes are pairs of

---

[2] Recall that the citric acid cycle is also known as the Krebs Cycle or Tricarboxylic Acid Cycle. The name used in your class will be the one favored by a particular textbook or instructor.

chromosomes that share a similar genetic pattern. If one chromosome has a gene affecting hair color positioned at nucleotide 40,005, then its homologous chromosome will also have a hair-coloring gene at nucleotide 40,005. The two genes aren't necessarily the same gene though. One might color the hair yellow, while the other might color it brown; they are two *versions* of the hair-coloring gene.

In genetics, we call different versions of the gene **alleles**. We call the position of the gene its **locus**. To recap: a gene is a physical stretch of nucleotides that codes for a protein. A locus is the *location* of the gene on the chromosome. Throughout a population of organisms, different versions of the genes can be found at that locus, and the *versions* are called alleles. Within the population as a whole, many more than two alleles could be floating around of course. For instance, one individual might have green and blue hair color alleles while another individual might have an allele for chartreuse and razzmatazz. Who knows? There are lots of versions of genes floating in the gene pool, but within one individual, you find a max of two for any given locus.

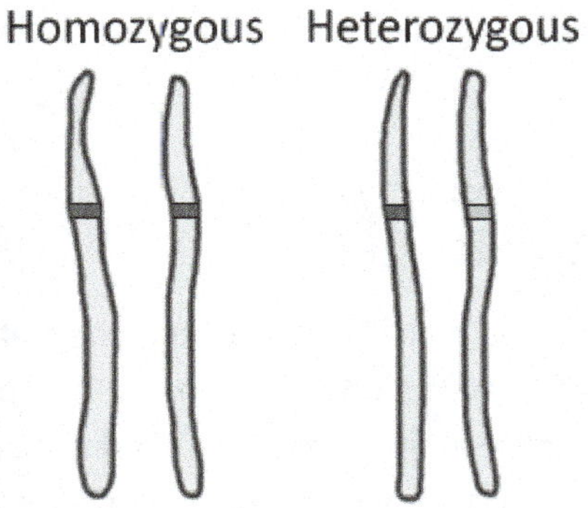

Suppose the two alleles at a location are identical, (both coding for blue hair, say). In that case we say the organism is **homozygous** at that locus. If the alleles are not identical, (one for blue hair and one for orange hair), then we say the organism is **heterozygous** at the hair color locus.

Let's adopt a sample critter to illustrate. Take the fictional creature *Mungabeastius biggus*. Let's say one particular mungabeast has two alleles for hair color: light gray and white. When the mungabeast makes gametes, each homologous chromosome goes its separate ways during meiosis. Fifty percent of the gametes will get the allele for light gray hair and the other 50% will get the allele for white hair. That means that 50% of the mungabeast's offspring will inherit the light gray allele while the other will inherit the white allele. This is called the **Principle of Segregation**, which is a fancy term that basically means that the two alleles of a locus will get segregated away from each other when it comes time to make gametes. Both the mother and father do this, and when mating occurs the homologous pair is reformed when each parent donates his or her version of that chromosome.

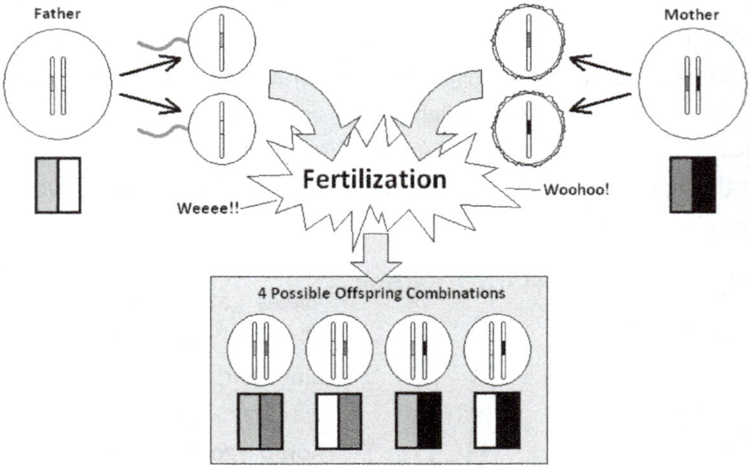

The trick is that when the gametes come together at fertilization there are a number of combinations that can result. To see how this is true, consider the image above. Suppose that dad has the aforementioned light grey and white alleles, whereas mom's two alleles at the same locus are one dark grey and one black. Now, when mom and dad contribute gametes during fertilization we find that there are four possible outcomes of this jumble. The first is light gray and dark gray. The second is white and dark gray. The third is light gray and black. The last is white and black. This genetic shuffle is at the heart of inheritance, and we'll see it over and over and over . . .

But let's back up to cover an interesting question about what happens during the time that the two alleles are within the same cell. Suppose an organism is heterozygous for hair color, like our mungabeasts. In other words, let's say it has one allele for black hair and one for white hair. Which version of the gene gets **expressed**? What color of hair will the organism actually *have*? To examine the answer, let's look at some basic patterns of inheritance that result as a consequence of the Principle of Segregation.

Blending Expression (or Partial Dominance)

The simplest resolution to which allele would lead to the organism's phenotype is that *sometimes* there is a stalemate between the two. Expression in this fashion is a blend of the two traits. For example, suppose that both mom and dad have the same alleles, say alleles that lead to blue or yellow pigment production. If both alleles created proteins that worked equivalently well, the combined color would end up being green (blue + yellow = green). Both mom and dad would have green hair. This is a simple blending, or **partial dominance**, as we say in biology, because both alleles assert themselves, each winning partially.

However, what will happen if they have offspring? Both parents could produce two types of gametes: those packaged with the blue allele, and those packaged with the yellow allele. Then, when you put the gametes together, you have four different ways of doing so. One offspring could get both blue alleles, two offspring could get one blue and one yellow, and one offspring could get both yellow alleles. This event is typically represented by drawing a **Punnett square** (next page).

If you examine the Punnett square, the first thing to note is that three different phenotypes emerge. The blue/blue (bb) genotype

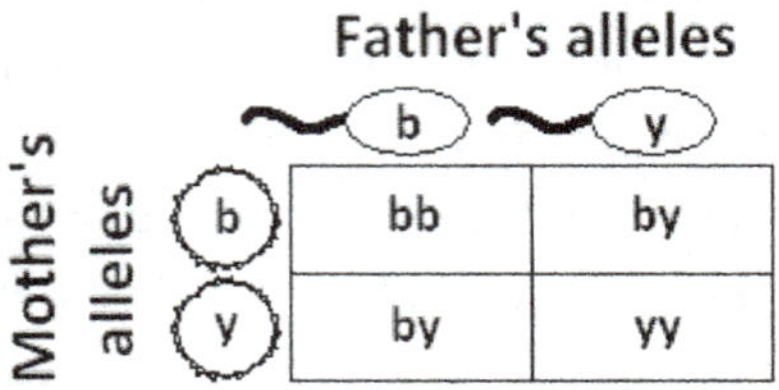

leads to a blue haired phenotype. The blue/yellow (by) genotype leads to a green haired phenotype, and finally the yellow/yellow (yy) genotype leads to a yellow haired phenotype. The distribution of these phenotypes is in the ratio of 1:2:1, meaning if the parents had four offspring, the odds are that one would have blue hair, one would have yellow hair, and two offspring would have green hair. This is the typical pattern in partial dominance inheritance.

Dominant/Recessive Expression

There is another possible outcome to the struggle between two alleles, an outcome called **dominant and recessive expression**. The basis of this kind of allelic interaction is as follows: the function of the protein of one allele is almost unnoticeable compared to the function of the other. In other words, it seems as though one allele overpowers the other. For example, one allele might code for a protein that is simply defective; it has no function at all. If the competing allele makes a protein that has *any* function at all, it wins by default. Another possibility might be that one allele's protein completely negates the effects of the other allele. Regardless of the reasons behind it, the expression you see is very distinctive: anytime the two alleles are put together into an organism, one allele gets its way over the other. The phenotype that results is determined by the dominant allele. The other allele is silent.

In the image that follows, we see a mungabeast that is heterozygous for both hair and eye color genes. In the picture, the hair alleles behave with partial dominance so that the black and white alleles mix to make the grey hair phenotype. With eye color, the genes for the mungabeast behave with dominant/recessive nature so that only one of the alleles wins (in this case the grey one[3]). Note that the picture doesn't necessarily represent the way hair and eye colors work in other organisms. It's just meant to show how blending and dominant/recessive expressions work.

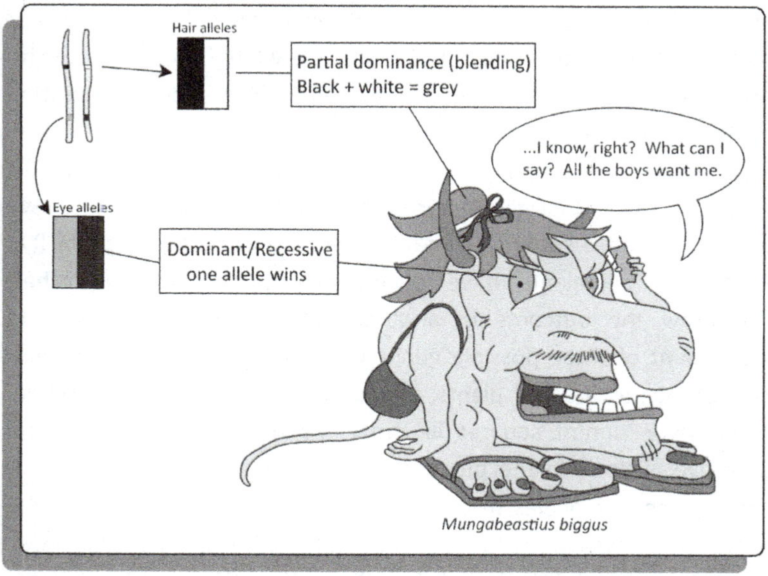

Mungabeastius biggus

Anyway, as it turns out, dominant/recessive expression has a great deal of historical importance in the study of inheritance, because it was the careful experiments to investigate this phenomenon that put the field of modern genetics on the map. Particularly significant were the experiments by **Gregor Mendel**, an Austrian monk living in the mid-nineteenth century, who used careful cross-mating of pea plants to draw important conclusions.

Gregor Mendel started with a pure strain of pea plants that always gave red flowers, and another pure strain that always produced white flowers. Based on our first experience with

---

[3] Hey, what do you expect? It's *grayscale* after all!

expression patterns, one would guess that by crossing them you would get a pink flowered pea plant. What Mendel found was that all the first generation offspring had red flowers. The white trait was nowhere to be seen. Strange! Moreover, when he crossed all these first offspring generation flowers with each other, three out of four made red flowers, and one out of four made white flowers. The white flower trait had somehow survived hidden in the first child generation of all red flowers!

We can decipher the mystery by drawing another Punnett square using the first generation as the cross (see below). When we do this, we get a second generation that all have a genotype of Rw. These are all red plants. If we now cross these plants with each other, and draw the outcome in another Punnett square, we see an assortment of three possible genotypes: RR, Rw, and ww. The first two genotypes are red plants, and only the ww genotype is white. Those two Punnett squares illustrate the nature of dominant and recessive alleles. In the presence of a dominant allele (here the R allele), recessive alleles (here the w allele) do not contribute to the overall phenotype.

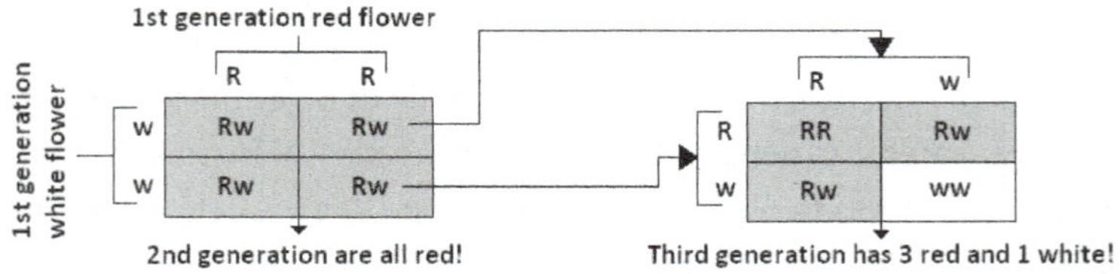

In conclusion, the pattern of inheritance that emerges from dominant / recessive gene expression is a 3:1 ratio.

Middle-of-the-Road Expression

On one end, you have the perfectly blended expression, and on the other end you have the dominant/recessive expression. Think of these as two extremes. One could easily imagine a situation that is midway between the two: one allele is highly outspoken over the other, but the other allele is not entirely silent. Suppose Mendel's pea plants had been like that. Instead of finding the ratio of three red and one white plant, you might find one red, two slightly-less-than-red, and one white. Without careful observation and measurement, you might simply conclude there were three red and one white and jot down the result as dominant and recessive.

The purpose of pointing out this middle-of-the-road pattern is that inheritance can be a complicated game to play, and deciphering which gene is doing what can be a monumental task. In any event, Punnett squares can help you see the underlying phenomenon. For any given pair of alleles, four types of offspring are possible, and depending on the degree to which the two alleles compete with each other, you can get a clear 3:1 ratio, or a 1:2:1 ratio, or a ratio that is deceptively close to 3:1. A sharp eye and careful measurement are crucial in biology!

Other Expression Patterns: Sex-Linked and Maternal Inheritance

Up to now we have only discussed homologous chromosomes. Technically, not all of a diploid's chromosomes are homologous. There are usually two chromosomes in the entire bunch that do not have a second copy. These are the **sex chromosomes**, designated "X" and "Y." All the others, (those we've talked about previously), are referred to as the **autosomal chromosomes**. A diploid organism will either have one **X chromosome** and one **Y chromosome**, or two X chromosomes. For most species, if an organism has two X chromosomes, it is female. If it has one X and one Y, it is a male.

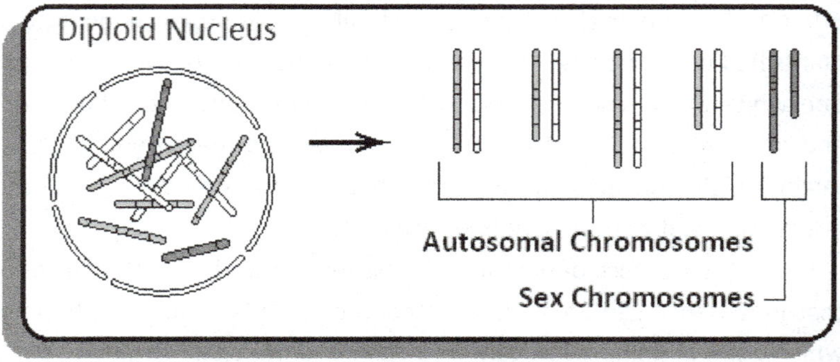

A unique phenomenon called **sex-linked inheritance** occurs when an allele pair is located on the sex chromosomes. Many of the gender-specific genes responsible for making us male or female are found on the X and Y chromosomes, and as a result, large parts of their respective DNA differ dramatically. Still, they are paired, but a locus on one may not have a homologous locus on the other. This latter fact means that some sex-chromosome alleles will not be paired with an equivalent partner like the alleles on the autosomal chromosomes. So, if a recessive gene appears on an X chromosome, and there is no equivalent locus on the Y chromosome to affect it, it will automatically be expressed in a male (since the recessive gene has no competition from an allele on the Y).

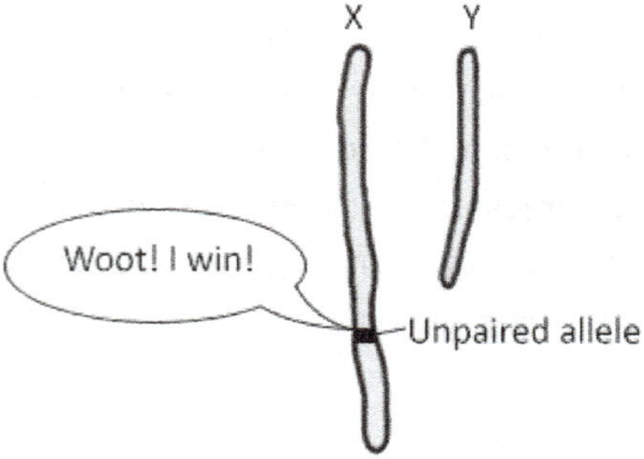

If, however, another X chromosome is present, such as in a female, the usual dominant / recessive or blending patterns will be seen.

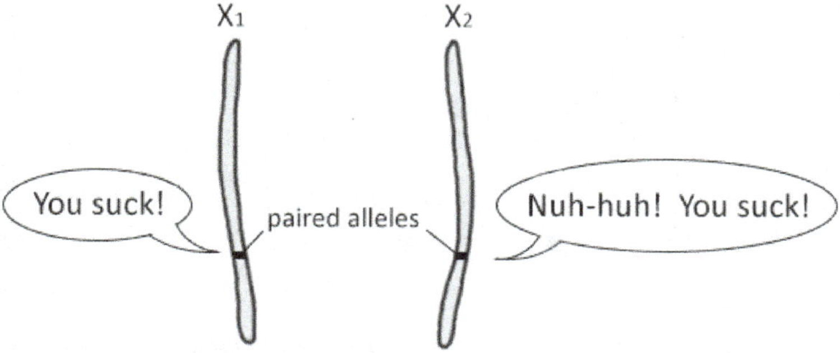

The important point is that the inheritance will be different for males and females. More often than not, males get the raw deal. Since they only get one copy of the X chromosome, any bad allele on that chromosome gets expressed. This harkens back to the argument that haploid organisms have less stable genomes. They only have one copy of each allele, and if that copy is not good, they're in for trouble. Males in that sense are partially haploid, you might say[4], at least in so far as their X chromosome is concerned, since many of the genes on that chromosome have no partner.

On the other hand, if there are alleles on the Y chromosome that are not present on the X chromosome, males once again have only the one copy to use. This gives them the raw deal again, because any bad genes appearing on the Y chromosome will be expressed.

---

[4] It is technically not correct to refer to males as haploids. I only point out this slight parallel just to illustrate how diploid genomes can be more stable than haploids.

The punch-line for sex-linked chromosomes is that these kinds of traits are expressed mostly in males, primarily because they only have one copy of the allele in question, either on the X chromosome or on the Y chromosome.

Lastly, there is one more pattern of inheritance that differs from the usual autosomal patterns we discussed at the beginning. This last pattern is **maternal inheritance**. As you may recall, the mitochondria (the powerhouses of the cell) have their own DNA, which may be left over from when they lived as independent bacteria-like organisms. Mitochondria are typically inherited from the mother (stored in the egg). This means that if a mitochondrion has a gene that gives rise to a very specific trait, all the offspring of the affected mother will inherit the mitochondrial gene, regardless of the father's genetics. This form of inheritance is generally easy to follow, because the trait will follow along the maternal lines of a family tree. Mothers will have children that are all affected. Affected daughters will have children who are all affected, and so forth. In contrast, the males of the family will never pass the trait to their offspring. For us males, this is a bit of a reprieve, since we normally get the brunt of the genetic attack.

Those are the basic inheritance patterns. In the next unit, we'll take some time to discuss pedigree charts, which uses pictures to show inheritance from one generation to the next. After we talk about pedigree charts, we'll take a look at what happens when we examine the patterns that result from *more* than one locus (hair color and eye color combined, for example).

Summary

In this section, we discussed the basic patterns of inheritance.

- A gene is a sequence of genetic information that codes for a protein. For any given gene, there are usually a number of different versions of it floating around in the population. Each version is called an allele. The location where the gene appears is called its locus.
- A creature's genotype is the set of alleles the creature has. Its phenotype is how those alleles have been expressed. The phenotypes that appear in the wild are called the wildtype.
- Homologous chromosomes that have the same allele (i.e. same version of a gene) at some specific locus are said to be homozygous at that locus. Homologous chromosomes that have different alleles for a given locus are said to be heterozygous at that locus.
- When a male and female make gametes, partner alleles on homologous chromosomes are sorted independently into different gametes. This is called the Principle of Segregation.
- If a male and female each have two different alleles for a given locus, then there are a maximum of four ways in which those alleles can be partnered in the offspring.
- In individuals that are heterozygous (two different alleles), a blending effect can sometimes be seen as the two alleles compete to be expressed. This type of inheritance pattern is called partial dominance, because each allele expresses some dominance. Offspring of heterozygous individuals of partial dominance alleles fall into ratios of 1:2:1 based on their phenotype.
- When one allele is completely dominant over another, the dominant allele will determine the phenotype of the organism. This type of inheritance pattern is called

dominant/recessive. This is a common pattern and is historically important. Offspring of heterozygous individuals of dominant/recessive alleles fall into ratios of 3:1.

- Some inheritance patterns fall in between the partial dominance and dominant/recessive patterns, so that the offspring appear to have a 3:1 ratio, when in reality a subtle difference leads to the 1:2:1 ratio.
- Most diploid organisms have two sex chromosomes, designated X and Y. Typically, the male are XY, and the females are XX. The non-sex-chromosomes are referred to as the autosomal chromosomes. The Y chromosome typically differs radically from the X chromosome.
- Sex-linked inheritance patterns occur when the allele producing a particular trait appears on one of the two sex chromosomes. In males, this often leads to an allele that has no partner, so that it always expresses its phenotype. In females, the allele in question will have a partner allele (since there are two X chromosomes) so that the phenotype is determined by the same mechanics as alleles on autosomal chromosomes.
- Females in most species provide the mitochondria for their offspring. The mitochondrion in the sperm of the father is broken down and not passed to offspring. Due to this, a special form of inheritance called maternal inheritance can result, where all the children of the mother inherit the mitochondrial trait in question, without exception.

Unit 2

Pedigree Charts

In this section we cover:

1) Pedigree Charts for:
2) Autosomal Dominant/Recessive Traits
3) Sex-Linked Traits
4) Maternal inheritance Traits

When we talk about inheritance, we often draw little diagrams that show how a trait travels from one generation to the next. These little diagrams are called **pedigree charts**, and a sample of one is as follows:

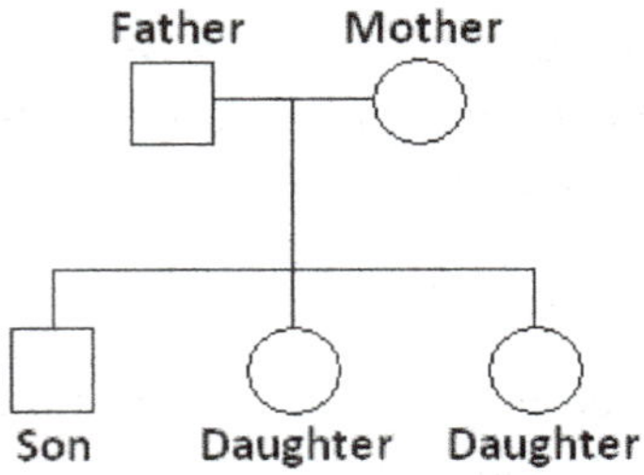

In this pedigree chart, we have two parents and three children. The first thing to notice is that males are designated as squares and females are designated as circles. When a pedigree chart is used to show how a trait travels in a family, the squares and circles are darkened in. For example, let's say we want to see which individuals in the family have freckles (a supposedly dominant/recessive trait in humans). In this case, we might see:

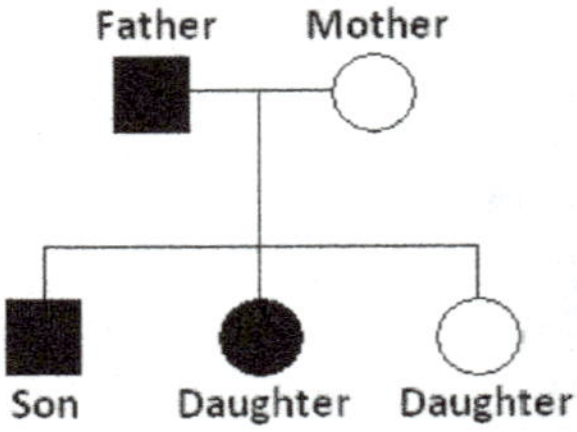

This chart means that the father and two of the three children have freckles. Dominant traits can usually be spotted in this manner. If a parent has a dominant trait, then half or more of the children will often have the trait as well (it can surprise you though). If one parent is homozygous for the trait (say homozygous for freckles, f/f), then all the children should end up with freckles.[5] Since less than all the children have freckles in this example, we can conclude that, (unless something else isn't involved to complicate things), the father is not homozygous dominant.

Pedigree charts are especially good at helping us figure out if a gene is dominant or recessive. Let's look at another chart and discuss the possibilities:

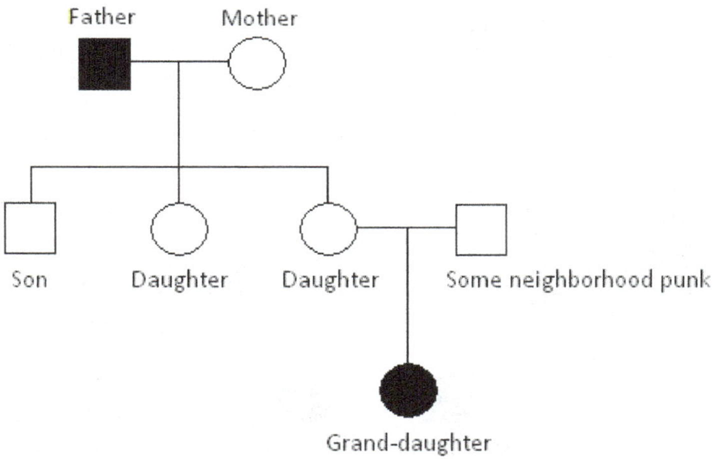

In this chart, we see that the father has some kind of trait. Say he has the inheritable disease "Goofadoofosis", and we see that the granddaughter has it as well. We can conclude from this chart that – in the absence of complicating factors – Goofadoofosis is not inherited in a dominant fashion. If it were, the second daughter would have had it as well. Since the second daughter does not have it, it must be inherited in some other fashion, possibly as a recessive trait.

---

[5] That is, of course, unless other genes are involved, in which case it gets a little more complex. We'll investigate this a little later.

At any rate, this is commonly what a pedigree chart will look like for recessive traits. The trait will appear periodically, then vanish and skip generations. Since it is recessive, it requires both parents of any given union to have the recessive allele, and if the allele is particularly rare in the population, the disease will skip many generations before appearing again. In contrast, as we saw earlier, dominant traits tend to have a clear path; every generation is affected.

The pedigree charts for sex-linked inheritance are very recognizable as well:

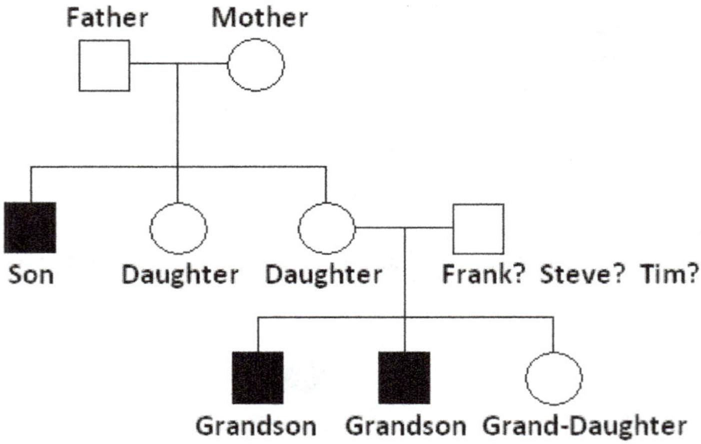

As discussed in the previous section, sex-linked inheritance tends to zap male offspring much more often than females, so the pedigrees for these kinds of traits will show an unusual concentration of affected males and little to no affected females. The most likely explanation for this particular pedigree chart is that the grandmother had the allele in question on one of her X chromosomes. The son inherited that X chromosome, as did the daughter on the far right. That daughter then gave the defective X chromosome to both her sons.

The pedigrees for maternal inheritance will typically show all affected individuals linked through the mothers:

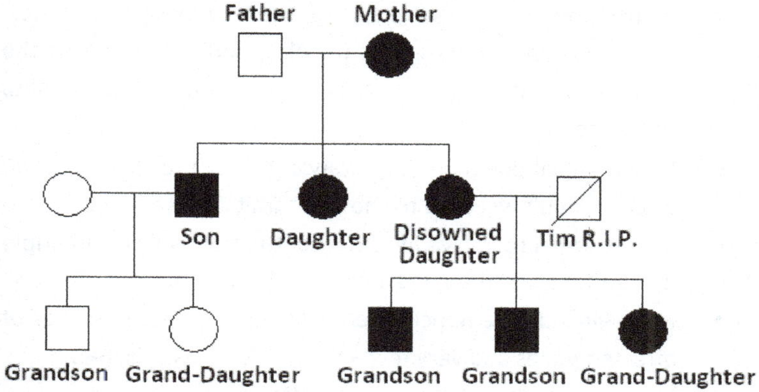

Note how the male on the left does *not* transmit the trait to the offspring, but all the females do. This is very common for the maternally inherited traits.

Lastly, individuals that are deceased are usually drawn with a slash through them (such as poor Tim in the above diagram).

Summary

In this section, we discussed basic pedigree charts.

- In pedigree charts, squares are used to designate males. Circles are used to designate females. Individuals having the trait in question are usually drawn with their symbol darkened in.
- In autosomal dominant inheritance, half of the children of an affected adult will tend to show the trait.
- In autosomal recessive inheritance, the trait will skip multiple generations and then unexpectedly show up again.
- In sex-linked inheritance, there will be an unusual number of affected males and very few – if any – females affected.
- In maternal inheritance, the pedigree chart will show all children of the affected mother as having the trait in question.
- Deceased members of a pedigree chart are often drawn with a slash through their symbol.

Unit 3

Two Loci Inheritance

In this section we cover:

1) Inheritance patterns of two loci on two homologous chromosome pairs
2) Inheritance patterns of two loci on one homologous chromosome pair
3) Effects of the distance between loci on cross-over probability

We are now in position to examine what happens with the inheritance pattern of two simultaneous loci. That is to say that we will be watching the inheritance pattern of two separate pairs of alleles.

When examining only 1 pair of alleles . . .

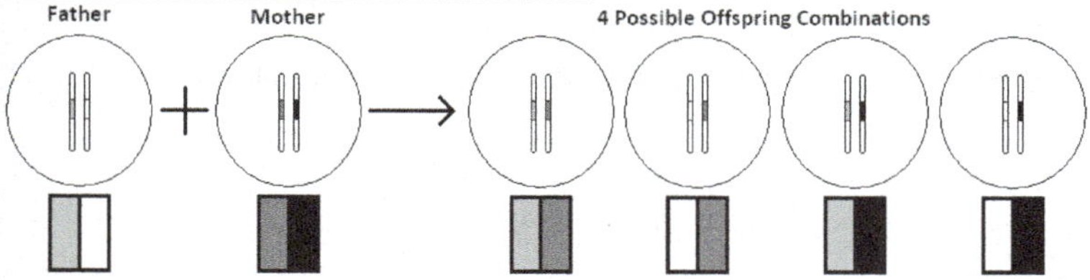

When examining 2 pairs of alleles at a time . . .

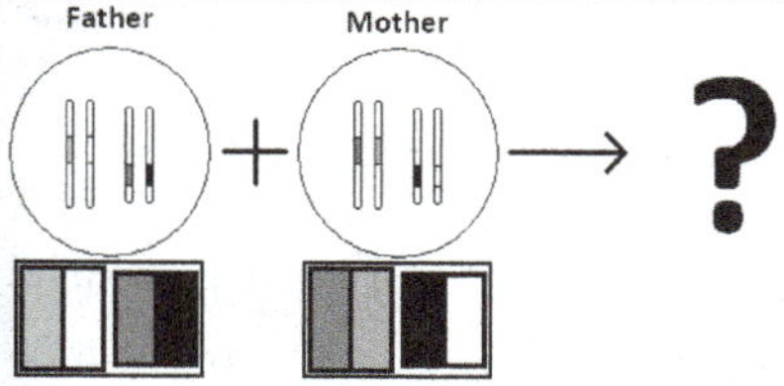

Why might studying this be important? Well, biological organisms contain more than one allele pair; they typically contain thousands upon thousands. Further, most traits that we see around us, (such as height, facial structures, diseases, and a slew of other things), are the result of a combination of many alleles working together as a group. The ultimate goal is to understand how these traits are passed from generation to generation, and the first step in that goal is to see how two pairs of alleles that each governs their own trait are inherited. In the next unit, we'll tackle how two pairs of alleles working together to make a single trait are inherited. Throughout this unit and the next, we'll be using a **three-step strategy**: 1) determine the types of gametes a genotype can produce, 2) put these together into a Punnett square, and 3) group the offspring genotypes based on their expression. The best way to understand this is to see it in action, so let's work through several examples.

Two Loci, Two Traits, Two Chromosome Pairs

When each allele pair is on a different pair of homologous chromosomes (shown below) each allele will segregate from the other. Let's consider our fictional mungabeast and examine two different traits: hair color and eye shape. Let's say that for hair color, it has one blue allele and one yellow allele, and for eye shape it has one allele for circular eyes, and one allele for triangular eyes.

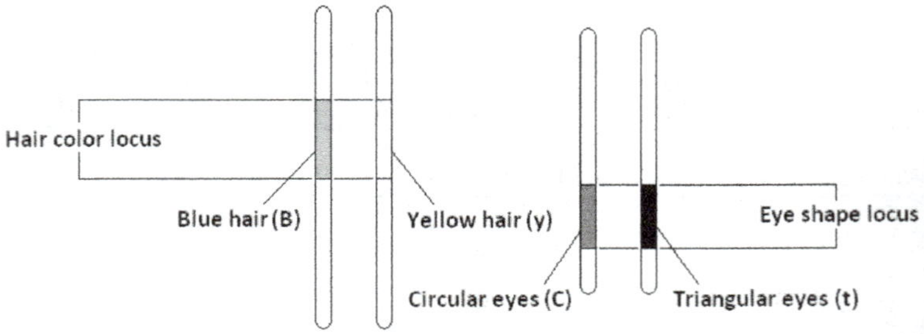

For shorthand, the blue allele will be "B" and the yellow allele will be "y." The circular allele will be "C" and the triangle allele will be "t." So, given this information, we know that the overall genotype of the organism is By/Ct. The first step in our strategy is to determine the types of gametes the organism can produce. When the mungabeast makes sex-cells, 25% of the gametes will have a blue allele and a circular allele (B/C), 25% the blue allele and triangle allele (B/t), 25% the yellow allele and circular allele (y/C), and the last 25% the yellow allele and triangle allele (y/t), as shown in the next image.

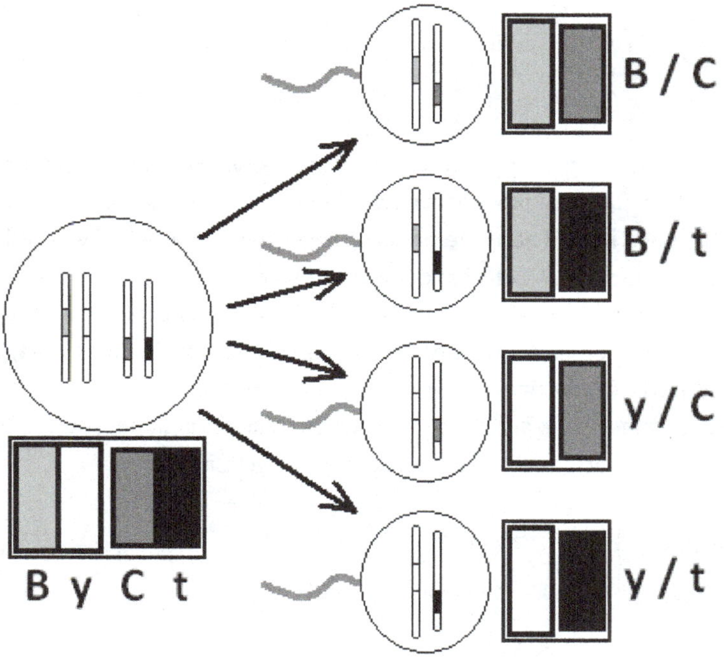

Now suppose this mungabeast mates with another mungabeast that has the same genotype (By/Ct). To determine the possible combinations in the offspring, we form the top and side of a double Punnett square with the possible gamete combinations. When we cross the gametes, we get a complex pattern with sixteen possible squares, each square being 6.25% likely to appear in the offspring. The resulting phenotypes (the color of hair and shapes of the eyes, etc.) depend on the individual expression patterns of the independent allele pairs.

	B/C	B/t	y/C	y/t
B/C	BB/CC	BB/Ct	yB/CC	yB/tC
B/t	BB/Ct	BB/tt	yB/Ct	yB/tt
y/C	By/CC	By/tC	yy/CC	yy/tC
y/t	By/Ct	By/tt	yy/Ct	yy/tt

→ 16 Possible offspring ↓ So, each is 1/16 likely ↓ 6.25% each

The last step is to see what phenotypes emerge. Let's say that both allele pairs behave with perfect blending (like the first pattern we looked at in the previous sections). What will the possible offspring look like? Let's look at just the hair color phenotype first and see. Twenty five percent of the offspring have two blue alleles, so they have blue hair. Twenty five percent have two yellow alleles, so they have yellow hair. Fifty percent have one blue and one yellow allele, so they have green hair. Looking at hair color alone, you see the same 1:2:1 pattern we saw in the previous section.

BB/CC	BB/Ct	yB/CC	yB/tC
BB/Ct	BB/tt	yB/Ct	yB/tt
By/CC	By/tC	yy/CC	yy/tC
By/Ct	By/tt	yy/Ct	yy/tt

Blue hair — Green hair — Green hair — Yellow hair

4 blue : 8 green : 4 yellow → 1 : 2 : 1

A similar examination will reveal the same pattern for eye-shape.

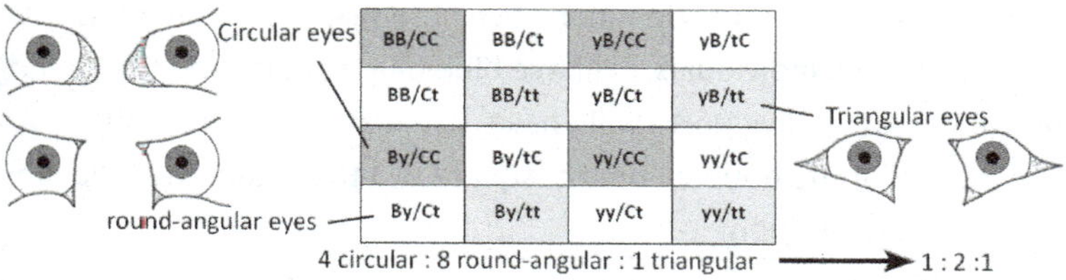

Let's use these results to see how many have, say, green hair and circular eyes. In order to have green hair, the mungabeast must have the By genotype, and in order to have circular eyes, it must have CC. So, to have green hair and circular eyes, it must have a genotype of By/CC. If you look at the double Punnett square, you'll see that two individuals meet that criterion. Since each of the two represents 6.25% probability, the total probability is 2 × 6.25% = 12.5%. Therefore, 12.5% of all the offspring will have *both* green hair *and* circular eyes. If we follow this approach for all the phenotypes, we arrive at the following table.

Phenotype	Probability
1 Blue, Circle (BB/CC)	6.25%
1 Yellow, Triangular (yy/tt)	6.25%
2 Green, Circle (By/CC)	12.5%
1 Yellow, Circle (yy/CC)	6.25%
2 Green, Triangular (By/tt)	12.5%
1 Blue, Triangular (BB/tt)	6.25%
2 Yellow, Round-angular (yy/Ct)	12.5%
2 Blue, Round-angular (BB/Ct)	12.5%
4 Green, Round-angular (By/Ct)	25%

From this list, you can see that when you have two pairs of alleles that each behave with blending expression, the total number of phenotypes is quite diverse and spread out.

Suppose, however, the two allele pairs behave closer to the dominant/recessive pattern of expression. Let's assume that the uppercase letters represent dominant traits; so blue (B) dominant over yellow (y), and circle (C) dominant over triangle (t). In that case, all the green haired individuals will have blue hair instead, and the round-angular eyed individuals will have circular eyes instead. An examination of the double Punnett square will then yield the following phenotype distribution.

Phenotype	Probability
9 Blue, Circle (B-/C-)	56.25%
3 Yellow, Circle (yy/C-)	18.75%
3 Blue, Triangle (B-/tt)	18.75%
1 Yellow, Triangle (yy/tt)	6.25%

The dashes in the table above indicate that it doesn't matter what the other allele is; the dominant allele gets its way regardless. To get an idea for how these numbers were found, consider the first entry, B-/C-. Nine of the sixteen will have at least one B *and* one C allele; these will be blue-haired *and* circular-eyed. To find the percent probability, we just divide nine by sixteen, which equals 56.25%. That means 56.25% of the offspring will have blue hair *and* circular eyes.

In that example, both parents had the identical genotype (By/Ct), but it would be easy to determine the outcome of any combination of parental genotypes. For example, if the father has, say, BB/Ct and the mother has By/tt, simply 1) list all the possible gametes that BB/Ct can produce and put them at the top of a double Punnett square and put all the possible gametes that By/tt can produce on the side. Then 2) fill in all the possible offspring in the sixteen given squares. Once that is done, 3) group the offspring genotypes based on how they compete for expression, be it blending, dominant/recessive, or anything in the middle range.

That said, let's look and see what happens if both allele pairs are located on the *same* homologous chromosome pair. Our overall three-step strategy will stay the same, but as we'll see, the major difference that will occur will be a large shift in the probabilities away from the base 6.25% probability for each square.

Two Loci, Two Traits, One Chromosome Pair

When the loci were on *different* chromosome pairs, each individual allele went its own way so that 25% of the gametes had alleles B/C, 25% had B/t, 25% had y/C, and 25% had y/t. Now, we look at two loci stuck on one chromosome pair.

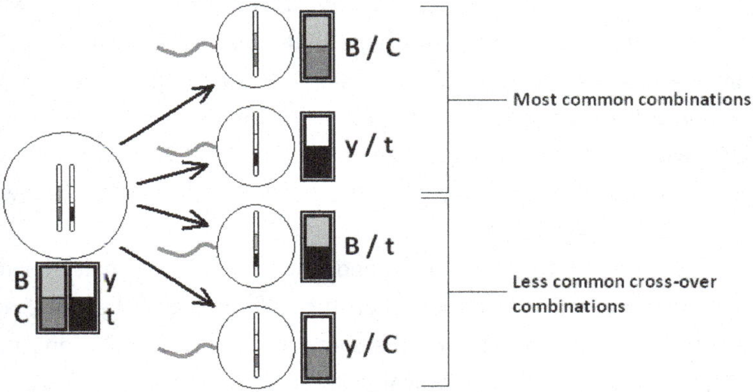

When the two loci are on the same chromosome, the alleles tend to segregate together into the same gametes. Most of the gametes that result are of one of two types: B/C, and y/t. A small percent switch partners during cross over to result in a small population of B/t and y/C gametes. Let's ignore this small fraction for the moment.

Suppose the mungabeast having this genotype meets another mungabeast with the same genetic makeup. To determine the possible offspring outcomes, we once again make a Punnett square. On the top we put the two most common combinations, B/C and y/t. On the side we put the same thing (since we've said that both parents have the same genetic makeup). As is usual with a simple Punnett square, we have four possible outcomes, as shown here.

Mother's alleles	Father's alleles	B/C	y/t
	B/C	BB/CC	yB/tC
	y/t	By/Ct	yy/tt

The yB/tC and the By/Ct are the same thing,[6] so if we compare the ratios of the offspring types, we get a 1:2:1 genotype pattern: one BB/CC, two By/Ct, one yy/tt. What will the phenotypes be? To determine that, we examine how each individual allele competes for expression. If they both behave with blending expression, you'll have a 1:2:1 phenotype distribution. The BB/CC will be blue-haired and circular-eyed. The two By/Ct will be green-haired with round-angular eyes, and the yy/tt will be yellow-haired with triangular eyes.

If, on the other hand, both allele pairs use dominant/recessive inheritance, then the By's will look the same as the BB's, and the Ct's will look the same as the CC's. You'll end up with the 3:1 pattern that is typical of the dominant/recessive type of inheritance: three blue-haired, circular-eyed individuals, and one yellow-haired, triangular-eyed individual (the one with the yy/tt).

A similar argument could be made if one pair behaves with dominant/recessive expression and the other with blending expression. Simply add up each phenotype and compare the ratios.

So far that's not unlike the genetics of a single allele pair, but what about the tiny fraction of gametes that resulted from crossing-over? Well, to be thorough, once you consider the possibility of crossing-over, then you are once again back to 4 different types of gametes that can be produced: B/C, B/t, y/C, and y/t (just like in the first half of this unit!). The difference is in the *probabilities* of each gamete combination. Previously, each gamete had a probability of 25%; each was equally likely. Now we are facing the situation of *two*

[6] The order in which we write the letters makes no difference.

loci on a *single* pair of homologous chromosomes, and the probability of the gametes having the cross-over products (B/t and y/C) will *not* be the same as the probability of having B/C and y/t. In fact, the chance of having a cross-over will depend on the distance between the two loci. The farther apart the two loci are, the higher the probability that cross-over will occur between them, and the less rare the B/t and y/C will be.

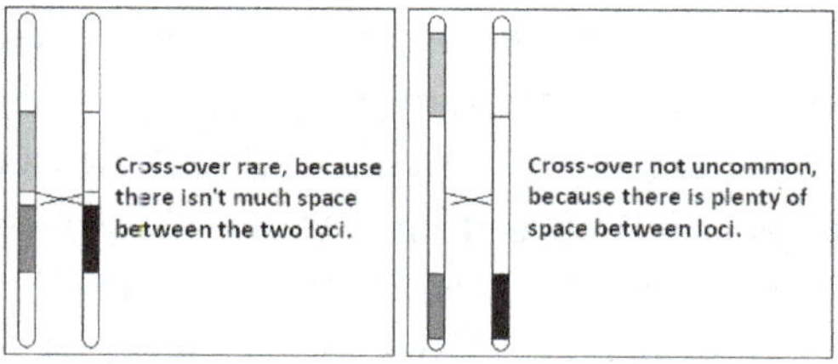

CONCLUSION:
The closer two genes are together, the more likely they will be inherited together.
The farther apart they are, the greater the chance they will segregate independently into different sex-cells.

Suppose the two loci are closely spaced to one another. In that case, it would be rare for cross-over to occur; B will almost always be packaged into the same gamete as C, and y will almost always be packaged with t. Only a small percentage of the gametes will have B packaged with t, and y packaged with C. As such, you might see the following probabilities for the gametes: 47.5% B/C, 47.5% y/t, 2.5% B/t, and 2.5% y/C[7]. Alternately, if the two loci are widely separated from each other, it might not be so rare that B crosses over to travel with t, or that y crosses over to travel with C. In that case, you might see a probability spread like this: 30% B/C, 30% y/t, 20% B/t, 20% y/C.

Regardless, this difference in gamete probabilities is one way in which geneticists can get an estimate of how far apart two loci are from each other. The farther apart they are, the more independently they are segregated into gametes, and the more independently they are inherited in offspring. If two alleles seem to travel together most

---

[7] These are just hypothetical numbers. They were not found using any formula.

of the time, we can conclude that they are closely spaced on the same chromosome.

How does the possibility of cross over affect the overall distribution of genotypes and phenotypes? Well, since we technically have 4 types of gametes, we need a double Punnett square. That, of course, means sixteen different offspring genotypes are possible. However, since the four different gamete types are not evenly distributed, the sixteen different genotypes are not evenly distributed either! When the four different gametes were each 25% likely, the sixteen resulting genotypes were each 6.25% probable. Now, the gamete distribution is skewed, so you end up with heavily skewed offspring genotype probabilities. For the sake of clarity, I'll draw out the table of probabilities that would result from a probability of 45% B/C, 45% y/t, 5% B/t and 5% y/C. Also, to minimize complexity, let's just assume that the two parents have the same genotype.

	45% B/C	45% y/t	5% B/t	5% y/C
45% B/C	20.25% BB/CC	20.25% yB/tC	2.25% BB/tC	2.25% yB/CC
45% y/t	20.25% By/Ct	20.25% yy/tt	2.25% By/tt	2.25% yy/Ct
5% B/t	2.25% BB/Ct	2.25% yB/tt	0.25% BB/tt	0.25% yB/Ct
5% y/C	2.25% By/CC	2.25% yy/tC	0.25% By/tC	0.25% yy/CC

This double Punnett square lists the probabilities of each individual genotype. How did I find these numbers? Well, to determine the combined probability of two independent probabilities, you multiply them together. So, the combined probability of 45% and 45% is:

$$0.45 \times 0.45 = 0.2025 = 20.25\%$$

Why did I write 0.45 and not 45% in the equation above? The reason is that 0.45 is another way of writing 45%. They are equivalent values, just different notation.

Several of these squares have the same genotype in them. For example, yB/tC and By/Ct are the same thing, so together their combined probability is 20.25% + 20.25% = 40.5%. Analyzing the rest of the genotypes in this manner we arrive at the following table:

Genotype	Percent	Out of 10,000 offspring
BB/CC	20.25%	2,025
BB/Ct	41.00%	4,100
yy/tt	20.25%	2,025
BB/Ct	4.50%	450
By/tt	4.50%	450
By/CC	4.50%	450
yy/Ct	4.50%	450
BB/tt	0.25%	25
yy/CC	0.25%	25

The last column represents how many offspring out of 10,000 offspring one might theoretically find to have the given genotype. So, if you had a brood of 10,000 children, like several people I grew up with, then you could expect roughly 2025 to have the BB/CC genotype.

Just as before, the next step is to break this down into phenotype distribution, which requires analyzing how each individual combination gets expressed. Let's suppose that the blue/yellow alleles behave in dominant/recessive pattern, whereas the circular/triangular alleles behave with blending pattern. That means all the By genotypes will look the same as BB genotypes. The CC's, the Ct's, and the tt's will each reflect their own phenotype. The table we just made will collapse into the following:

Genotype*	Percent	Phenotype	Out of 10,000 offspring
B-/CC	24.75%	Blue hair, round-eyes	2,475
B-/Ct	45.50%	Blue hair, round-angular eyes	4,550
yy/tt	20.25%	Yellow hair, triangle eyes	2,025
B-/tt	4.75%	Blue hair, triangle eyes	475
yy/Ct	4.50%	Yellow hair, round-angular eyes	450
yy/CC	0.25%	Yellow hair, triangle eyes	25
		* B dominant over y	

Once again the dash means the other allele is irrelevant. This table also gives us an idea of how wildly disproportionate some of the phenotypes can be. For example, the odds are that to find a yellow-haired, triangle-eyed offspring, the parents would have to have several thousand children before even *one* is born.

Let's take one more example of the above genotype frequency and consider what the phenotype frequencies would look like if BOTH allele pairs behaved in dominant/recessive pattern. In that case, all Ct's would look like CC's, so the phenotype frequencies would collapse into:

Genotype*	Percent	Phenotype	Out of 10,000 offspring
B-/C-	70.25%	Blue hair, round-eyes	7,025
B-/tt	4.75%	Blue hair, triangle eyes	475
yy/C-	4.75%	Yellow hair, triangle eyes	475
yy/tt	20.25%	Yellow hair, triangle eyes	2,025
		* B and C dominant over y and t	

Notice how close the pattern comes to the 3:1 pattern (B-/C- close to 75% and yy/tt close to 25%). That is the effect of having the two allele pairs on the same chromosome pair. The closer they are on the chromosome, the smaller their chance of being separated from each other into different gametes. As their chance of separation gets smaller, the 4.75% gets smaller as well. Suppose the two loci are close enough that the phenotype probability of B-/tt and yy/C- are each, say, 0.001%. In that case, you would have a table like this:

Two alleles positioned really close on one chromosome		
Genotype	Percent	Out of 100,000 offspring
B-/C-	74.999%	75,999
B-/tt	0.001%	1
yy/C-	0.001%	1
yy/tt	24.999%	24,999

From this table, you can hopefully see that the closer two alleles get, the more they begin to behave as though they were at a single locus. In this particular example, you'd need 100,000 offspring before the statistical outcome of one cross-over event happens to show the true nature of the individual loci. For some reason, I always seem to live next to the family that tries to do this.

So as the alleles get closer together, they begin to behave more and more like a single allele pair, but a similar argument could be made about their behavior when they are positioned farther and farther away. In the latter case, the more distance there is between them, the more independent they are. The more independent they are, the more they behave like two loci on two *different* chromosome pairs. It's a spectrum of inheritance patterns.

Single allele pair <---> two closely spaced allele pairs <---> two widely spaced pairs <---> two allele pairs on two chromosomes pairs

Summary

In this section, we discussed the inheritance patterns of two loci.

- When two loci are located on two separate homologous chromosome pairs, four possible gametes can result, each with 25% probability. When two such individuals mate, these 4 gamete types can lead to sixteen possible offspring, each 6.25% likely. The combined result is the same as a statistical superposition of two independent allele pairs.
- The phenotype distribution that results from sixteen different genotypes depends on the expression of the individual allele pairs. If both allele pairs behave in dominant/recessive fashion then the phenotype distribution will collapse into a 9 : 3 : 3 : 1 spread. If both allele pairs behave in a blending fashion, then there will be a large spread of intermediate phenotypes.
- When two loci are located on a single homologous chromosome pair, the individual alleles will tend to be inherited together. This results in two common gametes, along with two fewer common cross-over gametes. When two such individuals mate, the combined genotype distribution will be skewed away from the usual 6.25% probability per square. As before, the phenotype distribution will depend on the degree of dominant/recessive and blending tendencies.
- The closer two different loci are to each other on the same chromosome, the less likely will be a cross-over event. Two closely spaced loci will tend to produce an inheritance pattern similar to a single locus, whereas two widely spaced loci will tend to produce a pattern more like that of two loci on two different chromosomes. The degree to which two loci behave similar to a single locus gives us an idea of how far apart they are on the chromosome.

Unit 4

Polygenic and Complementary Traits

In this section we cover:

1) Polygenic trait inheritance

2) Complementary trait inheritance

In the last unit, we explored the way in which genotype and phenotype distributions occur in the inheritance of two allele pairs, each one governing a different trait. During that exploration, we used the three-step strategy: 1) determine gamete probabilities, 2) fill in the Punnett square, and 3) group the genotypes by expression. Now, let's wade a little deeper into the gene pool by studying the situation in which *two* allele pairs govern a *single* trait. A trait governed by more than one allele pair is called a **polygenic trait** because it requires more than one gene for its expression. Most human features are controlled by multiple genes . . . facial features, height, etc.

At any rate, in this unit we'll look at the length of the noses of the mungabeasts and pretend it is controlled by two different sets of alleles: locus A and locus B. In addition, we'll take a look at a unique distribution that arises when the trait requires at least one dominant gene at each locus.

Inheritance of Polygenic Traits

Let's look at an example of how these traits pass from one generation to the next. If the loci that control the single trait are located on different chromosomes, then the two pairs separate independently into different gametes, creating four possible gametes of equal likelihood, just as we saw in previous situations.

As before, if two such individuals mate then sixteen possible genotypes can result in the offspring. Since each gamete is 25% likely, each one of the sixteen offspring is 6.25% likely ($0.25 \times 0.25 = 0.0625$).

	A/B	A/b	a/B	a/b
A/B	AA/BB	AA/bB	Aa/BB	Aa/bB
A/b	AA/Bb	AA/bb	Aa/Bb	Aa/bb
a/B	Aa/BB	Aa/bB	aa/BB	aa/bB
a/b	Aa/Bb	Aa/bb	aa/Bb	aa/bb

AA/bB is the same as AA/Bb, so we collect them together into a combined likelihood of 6.25% + 6.25% = 12.5%. Collecting the other genotypes in a similar manner we get the following list of genotype probabilities:

Genotype	Probability
1 AA/BB	6.25%
2 AA/Bb	12.50%
1 AA/bb	6.25%
2 Aa/BB	12.50%
4 Aa/Bb	25.00%
2 Aa/bb	12.50%
1 aa/BB	6.25%
2 aa/Bb	12.50%
1 aa/bb	6.25%

As with all the other examples, the phenotype distribution that results depends on the pattern of expression that the individual alleles follow. Suppose that both allele pairs follow a blending expression. For instance, let us say that AA contributes three inches to the nose length, Aa contributes two inches, and aa contributes one inch. Suppose locus B contributes similarly (BB contributes three inches to the nose length, Bb contributes two inches, and bb contributes one inch). Since both allele pairs work together to determine the nose length trait, the combined effect of genotype

AA/BB would be a nose of length 3 + 3 = 6 inches. The genotype Aa/bb would have a total nose length of 2 + 1 = 3 inches. Genotype aa/bb would have a total nose length of only 1 + 1 = 2 inches. These are only a few examples, but if we examine all the possible genotypes, we would find that many of them result in the same nose length, so we can group them as follows:

Genotype	Percent	Nose Length (in)
AA/BB	6.25%	6
AA/Bb or Aa/BB	25%	5
Aa/Bb or AA/bb or aa/BB	37.50%	4
Aa/bb or aa/Bb	25%	3
aa/bb	6.25%	2

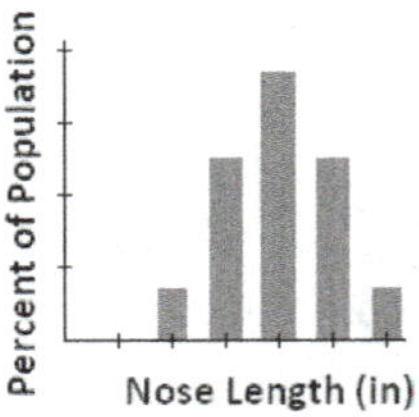

Suppose instead that the two loci contribute unevenly to the total nose length. For example, let's suppose that BB adds 5 inches, Bb adds 4 inches, and bb adds nothing. Let's say that locus A contributes as it did before, 3 inches, 2 inches, or 1 inch, depending on what alleles were present. The phenotype distribution would then look like this:

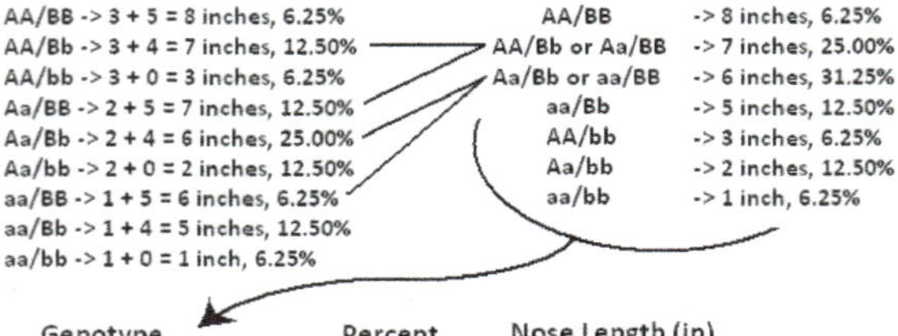

Genotype	Percent	Nose Length (in)
AA/BB	6.25%	8
AA/Bb or Aa/BB	25%	7
Aa/Bb or aa/BB	31.25%	6
aa/Bb	12.50%	5
AA/bb	6.25%	3
Aa/bb	12.50%	2
aa/bb	6.25%	1

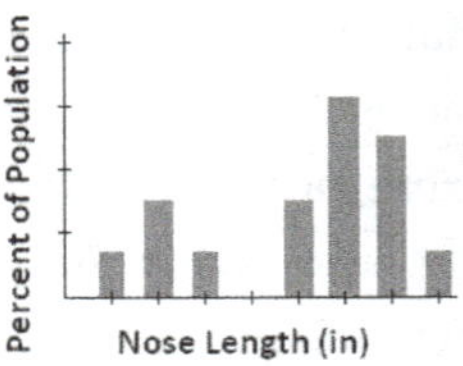

Let's do one more phenotype example from this. What happens if both loci behave in a dominant/recessive fashion? For instance, suppose that A is dominant over a: AA and Aa adds, say, 3

inches to the nose, whereas aa adds 1 inch. Let's say BB and Bb adds 3 inches and bb adds only 1 inch. The following phenotypes will then emerge (see graphic, following page).

AA/BB -> 3 + 3 = 6 inches, 6.25%
AA/Bb -> 3 + 3 = 6 inches, 12.50%
AA/bb -> 3 + 1 = 4 inches, 6.25%
Aa/BB -> 3 + 3 = 6 inches, 12.50%
Aa/Bb -> 3 + 3 = 6 inches, 25.00%
Aa/bb -> 3 + 1 = 4 inches, 12.50%
aa/BB -> 1 + 3 = 4 inches, 6.25%
aa/Bb -> 1 + 3 = 4 inches, 12.50%
aa/bb -> 1 + 1 = 2 inches, 6.25%

A-/B- -> 6 inches, 56.25%
A-/bb -> 4 inches, 18.75%
aa/B- -> 4 inches, 18.57%
aa/bb -> 2 inches, 6.25%

Genotype	Percent	Nose Length (in)
A-/B-	56.25%	6
A-/bb	18.75%	4
aa/B-	18.75%	4
aa/bb	6.25%	2

These look the same though!

56.25% : 18.75% : 18.75% : 6.25% ---> 9 : 3 : 3 : 1 ratio

Genotype	Percent	Nose Length (in)
A-/B-	56.25%	6
A-/bb or aa/B-	37.50%	4
aa/bb	6.25%	2

56.25% : 37.50% : 6.25% ---> 9 : 6 : 1 ratio

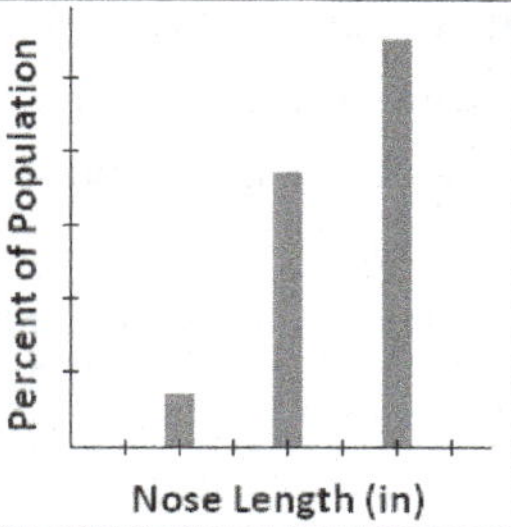

From the overall distribution, we see a 9 : 6 : 1 pattern emerge. This pattern is just the familiar 9 : 3 : 3 : 1 pattern in disguise – the same as the one we saw when examining the inheritance pattern of two dominant/recessive alleles pairs.

The punch-line for this whole business is that when two loci work together to affect one trait, you end up with a funky spectrum of phenotypes. In the first case, the curve looks like a normal bell curve. When the two loci contribute unevenly to the trait, such as in the second case, you get a weird curve, and this particular example has two distinct regions: a large part of the populace will have big noses and a smaller part will have pathetic little noses. Finally, in the case of alleles that behave with dominant/recessive pattern, we saw the phenotypes collapse into a much smaller distribution, here 9 : 6 : 1.

In theory we could extend the argument to traits that are governed by three alleles, or four, or five, and so on. In those cases, we follow the same strategy. Figure out the number and probability of gametes (which can be a complicated), fill in the Punnett square, and then group the resulting genotypes by how they get expressed. In the case of three alleles, for example, the Punnett square would be 8 × 8, because a genotype of, say, Aa/Bb/Cc can be segregated in eight different ways: A/B/C, A/B/c, A/b/C, A/b/c, a/B/C, a/B/c, a/b/C, a/b/c. So, a mating of two individuals having this genotype would require a Punnett square with eight possible gametes on the top and side. The probabilities also get a little more complicated, but the process is otherwise identical. In the end, you would end up with a smoother distribution curve. The more alleles involved in governing a trait, the more the distribution looks like a smooth curve rather than discrete columns.

Polygenics and phenotype distribution

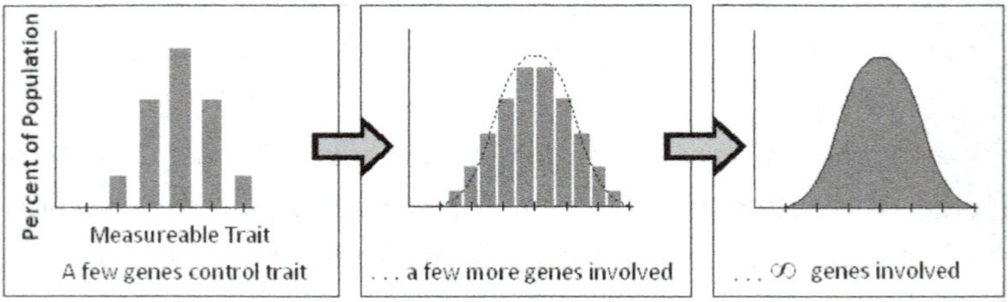

Finally, all these arguments can easily be expanded to cover the case of loci found on the same chromosome, as covered in the second half of the last unit. We use the same strategy: 1) find the gametes (here unevenly spread), 2) fill in the Punnett square, and 3) group the genotypes by expression. If the two alleles are close together on the chromosome, then most of the gametes will have them segregated together, and a small fraction will have them swapped due to cross-over. The overall phenotype distribution will be similar to that of a single locus, with a small fraction of rare cross-over phenotypes. If the two loci are far apart on the chromosome, they will be inherited closer to how two loci on two different chromosomes get inherited. How closely they approach the single locus pattern versus the two loci pattern gives us an idea of how far apart the two loci are on the chromosome.

Complementary Genes

In biology, it is not uncommon to encounter polygenic traits that require at least one dominant allele at each locus. These kinds of systems are said to be **complementary**. In the simplest case of two loci, A and B, there must be at least one A allele and one B allele or the trait fails to appear. In other words, the appearance of either aa or bb will cause the trait to fail. If a trait depended on three loci, A, B, and C, say, then there would have to be at least one A, one B, and one C present, or the trait fails. For example, if that system had Aa/bb/CC, it would fail, because it doesn't have at least one B. Important examples of these kinds of systems are most biochemical pathways, where each enzyme in the pathway is equally important. If even *one* enzyme in the pathway is coded by defective genes (say, aa instead of AA or Aa), then the *entire* pathway fails.

Let's see what would happen if we crossed an individual having a genotype of Aa/Bb with another having the same genotype. Once again, the gamete distribution and Punnett square is identical to that above, so we'll only concern ourselves with the phenotype distribution. After analyses, we have the following conclusions:

AA/BB -> Both enzymes function, 6.25%
AA/Bb -> Both enzymes function, 12.50%
AA/bb -> Pathway fails due to missing B enzyme 6.25%
Aa/BB -> Both enzymes function, 12.50%
Aa/Bb -> Both enzymes function, 25.00%
Aa/bb -> Pathway fails due to missing B enzyme, 12.50%
aa/BB -> Pathway fails due to missing A enzyme, 6.25%
aa/Bb -> Pathway fails due to missing A enzyme, 12.50%
aa/bb -> Pathway fails because both enzymes are missing, 6.25%

This collapses down into the following results:

A-/B-	-> Pathway works, 56.25%
A-/bb or aa/B- or aa/bb	-> Pathway fails, 43.75%
	56.25% : 43.75% --> 9 : 7 ratio.

In other words, the system works/fails in a 9:7 ratio. This important ratio is used to confirm whether a system of two genes is complementary.

The 9 to 7 Ratio of Complementary Genes

Summary

In this section, we discussed the inheritance of polygenic and complementary traits.

- When more than one gene contributes to a trait, the trait is said to be polygenic.
- A simple example of polygenic inheritance is when two alleles work in a dominant/recessive fashion to make a single trait. In that case, the distribution that results is 9 : 3 : 3 : 1, where the two 3's will look the same. This causes the 9 : 3 : 3 : 1 to collapse into a distribution of 9 : 6 : 1.
- If a polygenic trait relies on alleles that behave in a blending fashion, then the initial 9 : 6 : 1 pattern will spread out to a variety of curves. These can be bell-curved in shape, or can be funky with weird clumps and regions.
- The more genes that are involved in making a trait, the more the phenotype distribution will tend to spread out and get smooth, rather than appear as discreet columns.
- When a polygenic trait requires a dominant allele at each locus (i.e. at least one A, one B, one C, etc.), the alleles are said to be complementary. A system of two complementary genes will result in a phenotype distribution of 9 working : 7 failing. This is highly characteristic of complementary systems.

Unit 5

Other Genetics Terms

In this section we cover:

1) Epistasis
2) Penetrance
3) Expressivity
4) Modifier Genes
5) Pleiotropism

There are several more genetics terms that you will likely run into in a formal study of biology. Those terms are **epistasis, penetrance**, **expressivity**, **modifier gene**, and **pleiotropism**. I'll list and explain each.

Epistasis

Epistasis is to multiple gene pairs as dominant/recessive is to individual pairs. In other words, one locus can override a second locus, regardless of the second locus' genotype.

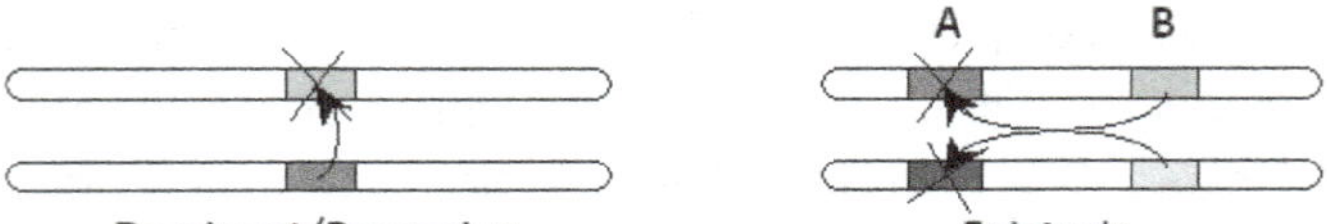

Dominant/Recessive

Epistasis

(The genes at locus B are "modifier genes" for locus A, see term below)

Let's show this by example. Suppose we are studying a single pair of alleles, one for brown hair (B) and one for white hair (w). Further, suppose that brown is dominant and we're crossing a population of parents that are all homozygous for brown (BB). To be clear with our setup: *not one of the parents has the white allele* – not a single incidence. Given that situation, there is no reason to suspect that a child with white hair should *ever* be produced. Right?

Suppose every several generations or so, a white-haired child is produced, nonetheless. Technically this should not be possible since we started with only B alleles. What's going on?! One explanation is that new mutations are taking place (essentially turning a brown hair gene into a white gene). That is possible but unlikely. Mutations are rare, so you would not expect it to be happening on a regular basis. Another explanation – and one that is much more likely – is that *another* set of alleles (at a different locus) are having an effect on the brown-haired phenotype. Thinking about it logically, one might expect that the expression of brown hair would depend on two factors: 1) the production of pigment and 2) the pigment's transportation to the hair. If the genes required for the transportation of the pigment fail, the brown pigment will not be transported to the hair. So, even though an individual may be homozygous for the brown allele (BB), it will still end up with white hair. The second locus – the transportation locus – overrides the hair color locus. That is epistasis.

The main lesson is this: if you cross two homozygous individuals and get something you totally did not expect, then the phenotype in question depends on more than just that one allele pair. Something *else* is also involved, and – in the present example - its effects are standing out.

Epistasis – The phenomenon where the phenotype of an allele pair is unexpectedly different because of the action of some other allele pair or pairs.

A real life example of epistasis is the appearance of albino mice in place of the usual gray or black color. The genes contributing to albinism are epistatic over the genes that create the color, so even though the appropriate color genes might be in the genome, the presence of the genes causing albinism overrides the color genes.

Penetrance

Penetrance is the percent of individuals who *should* have a certain phenotype (based on their genotype) and actually *do* have the phenotype. Let's continue the example we just used. Out of all the homozygous parents, one would not expect any children to have white hair – 100% should have brown hair. Suppose three out of every 100 individuals have white hair, for whatever reason. The penetrance for the brown hair phenotype is 97%. Penetrance of less than 100% indicates that there are more factors involved in the expression of a phenotype than just one allele pair. Epistasis is one reason, but environmental factors can play a part as well.

Penetrance is the percent of the phenotype that actually "penetrate" the populace as a trait.

Expressivity

In contrast to penetrance, **expressivity** refers to the variation of the phenotype itself; it's like the quality of the gene expression. For example, some individuals with "brown" hair might actually have light brown hair, while others might have dark brown hair. If the penetrance is 100% (i.e. no unexpected white-haired offspring) then you can measure the expressivity by dividing up the phenotype into the various groups. For example, you might measure the expressivity in terms of percentage – 30% have light brown hair, 60% have mid-tones of brown, and 10% have dark brown hair. Like penetrance, variable expressivity indicates that other factors go into the expression of the phenotype, most likely polygenics or environmental.

Expressivity is the variation of how the phenotype is expressed.

Modifier Genes

Modifier gene is a term that refers to the genes that alter the penetrance or expressivity of another gene. It's a generic term that can apply to most of the allele behavior we've talked about up to this point. For example, the "modifier genes" would be those genes that caused the 3% incidence of white hair in the population where you wouldn't expect white hair at all. "Modifier genes" might also be to blame for the variation in the color brown. You might expect an offspring to have dark-brown hair when it unexpectedly has light brown hair - something *modified* the color (assuming it wasn't something in the environment). When we talked about polygenic traits, any of the alleles in the group of genes that worked together to make a trait could be said to be a modifier gene of the other alleles. So, for example, if loci A and B both work together to contribute to the length of a snake, say, then we can say that the genes at loci B are modifier genes for the genes at A, and vice versa. If AA causes the snake to be one foot long, Aa causes it to be 3/4 foot long and aa makes it ½ foot long, then the genes at loci B "modify" that result by contributing their part to the overall length.

Note, however, that just because there is incomplete penetrance or variation in expressivity, it does not necessarily mean that other genes are to blame. There are lots of environmental factors that could go into altering the penetrance or expressivity of a gene, so just because some individuals have white hair or light brown hair, does not always means there are modifier genes doing it. After all, if you encountered an individual on the street with neon green hair, you would not assume they had a strange set of genes; they've obvious dyed their hair. Hair dye is an environmental complication, not a modifier gene.

A **modifier gene** is any gene that modifies the phenotype of another gene.

Pleiotropism

Pleiotropism is like the inverse of polygenics. Instead of a trait that is the outcome of multiple genes, it's a gene that causes multiple traits.

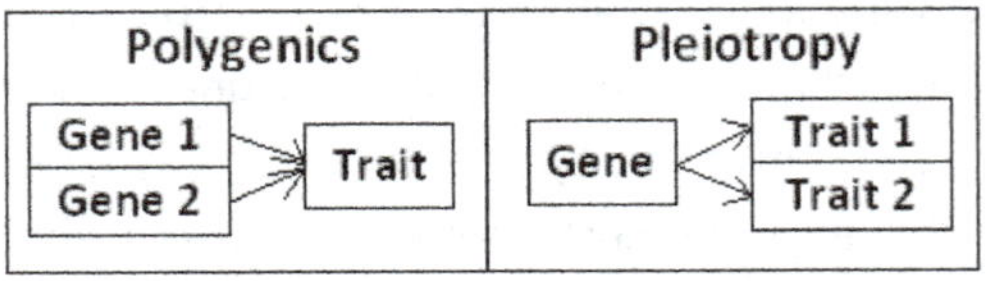

A classic example is from a series of experiments in the early twentieth century[8] involving the cross-mating of yellow and grey mice. Oddly enough, when a certain breed of heterozygous mice (Yy, "Y" for yellow fur and "g" for grey fur) were cross-mated, the result was a ratio of two yellow mice to one grey mouse. These results are unexpected. To see why, look at the following Punnett square:

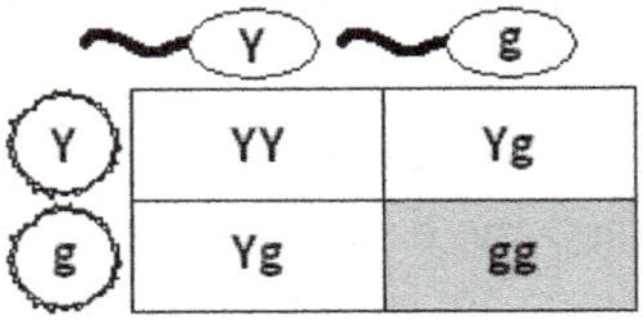

Expected result:

3 yellow
1 grey

From this Punnett square, you would expect a 3 to 1 ratio of yellow to grey offspring. The puzzle for the 2 : 1 ratio was solved by discovering that the gene in question, gene "Y", is a **lethal gene** to the mouse when homozygous: genotype YY is deadly. So, we would say that the allele Y is pleiotropic; it contributes two possible phenotypes: yellow hair *or* death before birth (when homozygous YY). Yes, in the broadest sense of the word, "dead" is a phenotype, and quite often its

---

[8] See the works of Lucien Cuénot.

inclusion in the results is the only way to make sense of a medical study.

Another example of pleiotropism is the gene leading to the human disease phenylketonuria. In that disease, a mutation in the gene that makes the protein phenylalanine hydroxylase leads to the inability to break down a chemical called phenylalanine. This chemical begins to react in other ways and build up as various side products that are poisonous. These side products lead to multiple traits: brain damage, heart defects, low-birth weight, abnormally small heads, and behavioral problems – to name a few. The disease is autosomal[9] recessive, which means that both alleles must have the mutation (or any mutation that makes the protein ineffective) in order for the disease to happen. Either way, it's clear that the mutated gene is pleiotropic: it causes multiple traits.

Pleiotropism is when one gene causes multiple traits. It's the inverse of polygenics.

[9] Recall that "autosomal" just means that the locus is on a non-sex chromosome.

Summary

In this unit, we discussed the various additional terms found in genetics.

- Epistasis is the process by which one pair of alleles overrides the phenotype of another, regardless of the other pair's genotype.
- Penetrance is a measure of the number of individuals in a population that have a genotype *and* its expected phenotype. If 100 individuals are expected to have brown hair, (based on their genotype), but only 90 have brown hair, then the penetrance of the trait is 90%. Penetrance less than 100% indicates some kind of complication, like another gene or environmental factor.
- Expressivity is the variation of a trait across a population. Variation in hair color from dark brown, to medium brown, to light brown is an example of expressivity. The presence of this variation is usually an indication that other factors are in play, such as other genes or environmental circumstances (up to and including the use of hair-dye).
- A modifier gene is any gene that modifies the phenotype of a second gene at a different locus. An example of a modifier gene would be one that alters the dark brown hair phenotype of another gene to light brown, or a gene that causes epistasis by shutting down the phenotype of another gene entirely. Remember, the modifier gene must be at a *different* locus than the gene it's modifying, otherwise it would just be considered a competing allele and be referred to in terms of dominant/recessive interaction.
- Pleiotropism is the phenomenon where one gene causes multiple traits. It is the inverse of polygenics.

Unit 6

Introduction to Evolution

In this section we cover:

1) Brief introduction to evolution, change and human philosophy on the matter
2) Fossil records
3) Radioisotope dating
4) Molecular and biochemical evidence of evolution
5) Morphological and embryological evidence of evolution
6) Biogeographical evidence of evolution
7) Microevolution and drug resistance

Up until now in this text and previous ones, we've studied all the basic chemical and structural processes of life, and we've examined some of the basic patterns in which traits are inherited. Now, we embark on a major topic biology: **evolution**.

In this unit, we're only going to give a qualitative definition of what evolution is and examine the evidence for it. In the next unit, we'll develop a more precise definition and then look deeper into the mechanisms that shape it.

What is Evolution?

In ancient times, mankind had a very static view of the Universe. We believed that all things never really changed. The planets, the stars, the continents, and even life . . . in the long run it stays constant. We believed that the Earth was at the center of all things and that everything else revolved around it. Humans, in turn, believed that they occupied an elevated status among living beings: we were just beneath the deities and far above the lowly animals. People believed that all things were created at the same time. Our view of the Universe was very anthropomorphic, (i.e., human-centered).

The path of science since those days has been a humbling one, as each of these beliefs has been shown to be highly unlikely, if not frankly incorrect. Not only have we discovered that the Earth is *not* at the center of the Universe,[10] but we've also found that we aren't above the animals at all. Everything about our biology works in nearly identical ways: we *are* animals. The only real significant difference, I might argue, is that we tend to think in more abstract ways. Further, millions upon millions of years of regularly layered fossils have suggested to us that not all species came into existence at the same time. On the contrary, the fossil record tells us a rather repetitive story; our version of life is only the most recent in a very long list of ever-changing sets of organisms. At the very start of the fossil record, we see life in very simple forms. The newer the fossil, the more complex it gets. After the dinosaurs, the slate seems to get mostly cleaned and the organisms get a little smaller, but are still complex. Overall, this gradual change in the structures of life is what we call evolution.

The basic idea behind evolution is rather simple: those organisms that have phenotypes that make them less able to survive and reproduce are less likely to leave offspring. Over time, the genotypes that control these less capable phenotypes become more

[10] The Earth is incredibly far from it in fact. As the great philosopher Douglas Adams put it, we're in a very unfashionable end of our galaxy.

and more rare, until eventually only the more "fit" genotypes remain. In other words, more often than not survival is for the "fittest" – which just means that those with the best genetics tend to have the best chance to survive and leave offspring. Then again, you don't necessarily have to be at the pinnacle of your species' genetics to have offspring, and in that sense evolution might be described as "survival of the fit-enough." Either way, change is introduced mainly in the form of 1) mutations in genes and 2) alterations in the environments, that both A) create new genotypes to be tested and B) alter the definition for what genotypes are best adapted. Essentially, mutations make new genes, and the ever changing environments are constantly changing their minds on what genes are best suited for survival. Over time, the species will gradually change. After millions and millions of years have passed, the species will no longer look or behave the same as it once did.

That's all well and good, but what evidence is there that this process is really taking place? Well, in truth, the growing list of evidence has become a veritable mountain. First, I need to reiterate the fossil record. Ever since men have been digging up old dinosaur bones, they have always found the fossils to be consistent with the layer in which they find them. They *never* find human bones mixed in with dinosaur bones. The more ancient the creature is the deeper the fossil will be. Dinosaurs, in turn, are *never* found among the remains of those that came before. This leaves a strong chronological record – you line up the remains and follow the dots.

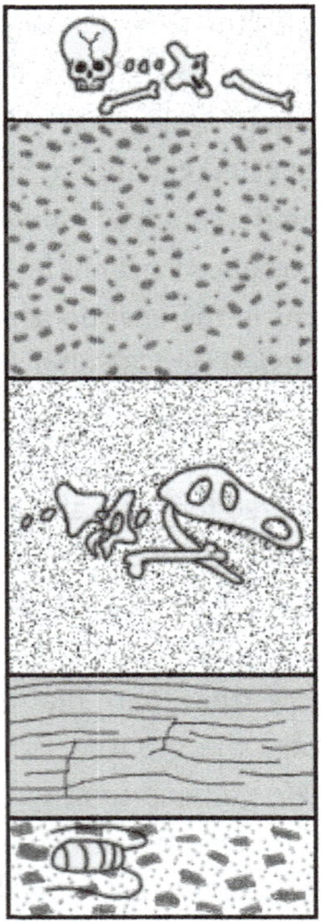

Fossils are always found in order of geological layers.

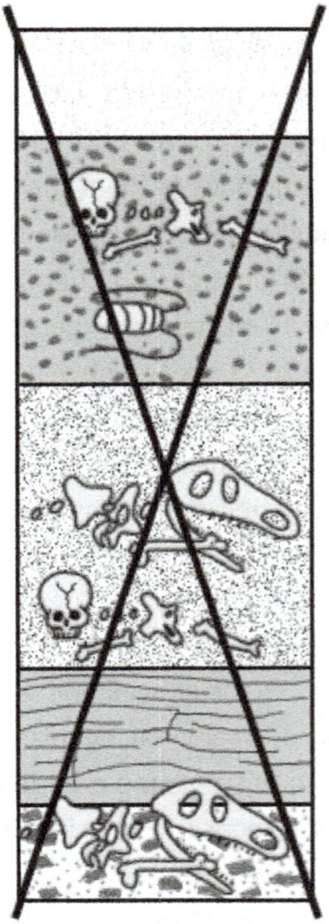

Fossils from different time periods are never found intermixed.

One may wonder whether this is just dumb luck, thinking that it's a matter of time before they find a fossil of a human and dinosaur together. Well, there is another way to date fossils other than by the geological layers in which they are found. One such way is **radioisotope dating**. Not to be confused with some kind of atomic couple's tango, radioisotope dating is a method by which the amount of nuclear radiation emitted from the substance is measured and used to estimate the age of the sample. In the early sections of Topic Text

A, we briefly discussed how some atomic nuclei are unstable. These particular nuclei often spit out pieces-parts in attempt to become more stable. One important example of this is carbon-14, an atom that has eight neutrons and six protons in its nucleus. It is formed on a regular basis in the upper atmosphere out of regular nitrogen (which has seven neutrons and seven protons) when the latter atom catches neutrons that constantly pelt the Earth from outer space. Once the nitrogen catches a neutron from the cosmic rays, it becomes unstable and spits out a proton to become carbon-14 (written ^{14}C).

$$\text{Neutron from outer space} + {}^{14}N \rightarrow {}^{14}C + \text{proton}$$

Carbon-14 is also unstable, but it breaks down slowly and (like all nuclei) at a very predictable rate. For any given sample of the stuff, precisely half will have broken back down into nitrogen after roughly 5370 years.

At any given moment, a small fraction of the carbon in the air is in the carbon-14 form. It gets formed in the upper atmosphere and breaks down again somewhere else, cycling through each form on a constant basis. Once plants take in the carbon-14 and combine it into glucose[11] (see the photosynthesis sections of Topic Text B) it becomes trapped and no longer has access to the upper atmosphere. That means the amount of carbon-14 in a sample of biological tissue will slowly drop over time since it cannot get "recharged" in the upper atmosphere. So, the older the sample is, the less radiation will come from it as the carbon-14 slowly breaks down. Given that, scientists measure the radioactivity remaining in a sample to get an estimate of how much carbon-14 is still left. Using this information, they can get a very reliable idea of how long ago the carbon got trapped. In a sense, the radioactivity acts like a clock for the age of the fossil. The image

[11] From there it enters into the biomass when eaten by other organisms.

below gives you an idea of how this works. The grey circles are the radioactive atoms. As time goes by, more and more break down into regular white circles. The number of grey circles that remains gives an idea of how old the sample is.

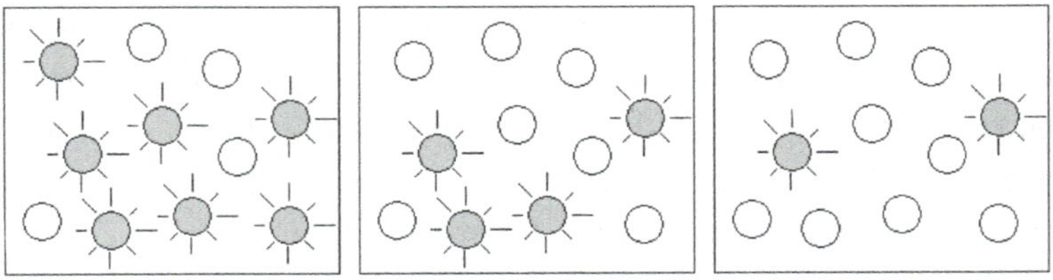

Young fossil **Medium-aged fossil** **Old fossil**

The point of the argument is simple: the fossil records always appear layered with the simplest forms of life at the most ancient levels. All other organisms are found at higher levels consistent with their fellow organisms. The fossils are never found to mix. Further, the radioactive dating of these fossils follows the same trend. The deepest fossils have the least radioactivity; they are very old. The dinosaur fossils have intermediate radioactivity; they are not as old. Human remains are always within, say, ten thousand years of age. Fossil remains of things that look *really close* to humans can be found before that. All of the radioisotope dating gives consistent results.

A consequence that immediately follows from radioisotope dating is that we are able to estimate the age of the Earth to be on the order of 4 to 5 billion years. This data is backed up by different radioisotope methods that use other kinds of radioactive nuclei, such as potassium-40, uranium, thorium, and so forth. This is in stark contrast to previous beliefs that the Earth is less than ten thousand years old.

Some of you may not be impressed with all this physics, nuclear and geological mumbo-jumbo, so let's look at some more evidence. Consider the biochemical and molecular information. With only minor differences, life all works by the same reactions and patterns. The main biochemical processes in you and me have been

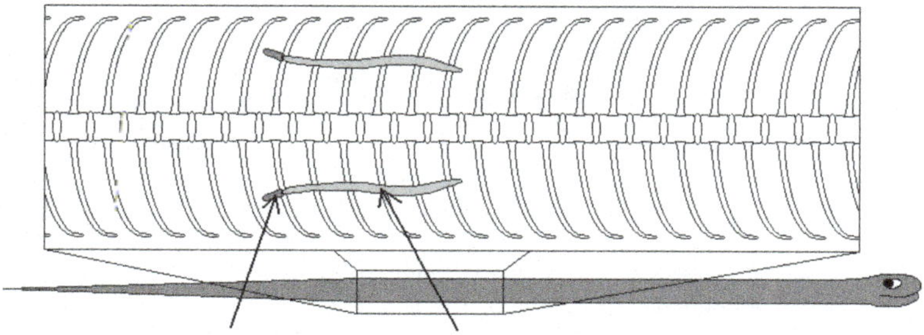

Why would a creature have hind-limb bones if it doesn't use them? The theory is that they evolved from primitive vertebrates that had limbs.[13] At some point, the ancestors of snakes and whales no longer needed legs at all, so now all that remains are tiny bones that are not used anymore. The fossil record is flush with all manner of intermediate body plans to support these ideas (i.e., snake-like creatures with smaller and smaller limbs, whale-like creatures with smaller and smaller hind-limbs, etc.).

Let's examine another area of evidence. Just after the egg and sperm combine during fertilization, the resulting fertilized cell undergoes a very dynamic period of growth and development called the **embryo** stage. While this is usually studied during the second semester of biology courses, (or even upper level courses), one observation from that stage gives us another clue about evolution. From what we can tell, all vertebrate animals go through the same initial stages; they look practically identical at first.

---

[13] The argument is something like this: The whale's ancestors lived along the coastal areas, and over millions and millions of years, their offspring that spent more and more time in the water were better suited for survival. Gradually, those that were suited to live in the water year-round became best suited for survival in that environment, up to the point that they stopped going on land. Now, one piece of evidence that they were once part of a land-walking species is the parts of the anatomy that still lingers but is no longer used.

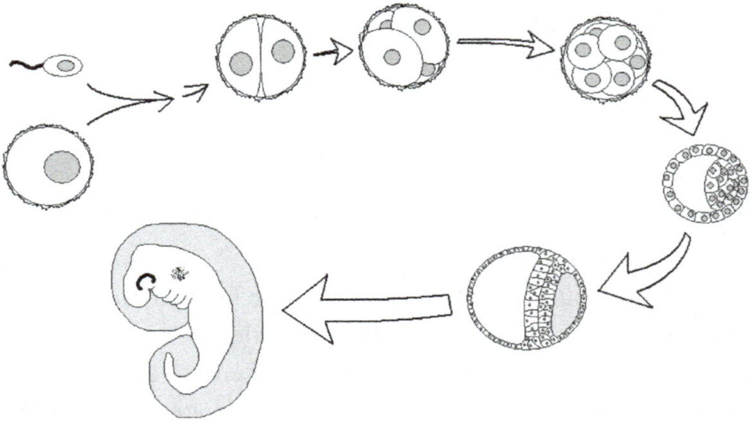

Fish? Bird? Shark? Whale? Bear? Cat?
Human? Weazel? Snake? Rat? Lawyer?

They all look like tiny aliens with tails. Even humans have tails at this stage, and the remnants of that tail stay with us as the tiny bones at the very bottom of your spinal columns . . . the ones just above your butt. Why would a human ever need to form a small tail? Why do all vertebrates start off looking practically the same? The prevailing explanation – based on countless fossil records, genetic evidence, geological studies, and other data – points to a common ancestor. As each new species developed from the original model, it carried with it the starting plan for development in the embryo stage. The more different two organisms are, the earlier they seem to depart from the original embryonic plan. For example, although fish start out looking the same as humans, they quickly diverge into their own growth patterns, whereas humans and chimpanzees look hauntingly similar for much longer in embryonic growth.

Let's go even further, this time examining **biogeographic** evidence. If species evolved from previous species, we would expect each geographic area of the world to have its own unique flavor of organisms, assuming humans hadn't tampered with the mix. Further, we would expect local areas to often have numerous variants of species that are closely related (e.g., hundreds of local kinds of butterflies, or beetles, or birds, etc.). In the cases where populations of creatures are separated from each other by long-term barriers, (such as an ocean), we would expect to see large variations between them.[14] The longer the separation, the more different the organisms might be.

Based on the geological evidence for continental drift, the major continents were all a part of one big continent approximately 200-250 million years ago[15]. Since then, the oceans have begun to slowly divide them, forming barriers for the movement of land animals. So do we find large differences in the animal populations from each of the continents? The answer, by and large, is yes. Each major land mass has a number of species that are specific to it, and most of these species have numerous local variants. If, however, you dig down far enough on more than one continent, you'll likely find fossils of the same ancestor – from a time when the continents where still connected. Thus we could imagine that when the continents parted the ancestor on one land evolved one way, while the ancestor on the other land evolved a different way.

Still, there are animals that seem to be widespread across the globe. Consider frogs. They appear in all the major continents except Antarctica. Frogs come in a vast array of colors and sizes. Does this counteract our biogeographic prediction? Not at all! The frog's design is actually very old and primitive. Early frog fossils date back as far as 200+ million years. Given their diversity, small size, and aquatic nature, it's not at all surprising that they are found just about everywhere. Further, it must be kept in mind that the same *kinds* of frogs are not found on each continent (again assuming humans

---

[14] The exception to this would be animals capable of getting over the barriers, such as most ocean-going organisms, many birds, and anything transported by humans, such as horses.

[15] This uber-continent is often called Pangea.

haven't messed with things). For example, South American frogs are not found in China, or Russia, or England, or any other place other than South America, and vice versa, unless *we* import them there.

Bigger animals, such as koala bears, kangaroos, giraffes, and buffalo have a much harder time getting from point A to point B. Once you put a barrier like an ocean in front of them, they're completely stuck. They stay on the same continent and diversify into a number of different breeds and species . . . until humans come along and muck everything up, that is. Of all our strengths and talents, our ability to completely screw things up is our paramount skill; try not to ignore the effects of humans on the world, especially when thinking about topics like evolution. After all, we didn't always have shopping malls, planes, boats, roads and so forth.

Another key piece of evidence to present about the existence of evolution concerns the **microevolution** of viruses and bacteria. We see this every day in laboratories and fight against it in an endless battle in hospitals. Viruses and bacteria are constantly experiencing random mutations, and their rapid reproductive cycle acts to speed up the evolutionary process, testing each new generation against the antiviral and antibiotics we use. Every once in a while, a new mutation occurs that allows the virus or bacteria to be immune to the current anti-pathogen medicine. This new super bug gives rise to others that are like it, and before long the entire population is replaced by organisms that are no longer affected by the drug. This is evolution in action, made obvious because of the extremely rapid rate at which reproduction and mutations occur.

Most of the rest of the biological world does not reproduce quite so rapidly, but the mutations are still happening inside all other organisms just as they are in bacteria and viruses. Mutations are the driving force for trying out new genetic formula, and over the course of millions of years, these mutations slowly accumulate, causing gradual unperceivable change between one generation and the next.

Arms slowly get longer or shorter. The shapes of eyes change, and ears get bigger or smaller. Tails get longer or vanish entirely. Flaps of skin attached to arms that help one generation fall gracefully from a tree get slowly larger until many generations later those who fall from the tree glide for long distances. Eventually, these flaps of skin become wings, allowing the descendants to fly from one place to place. This is the way of things in biology: constant change. Like the changing Universe revolving around something *other* than us, the biological world is constantly on the move.

Summary

In this section, we discussed the basic ideas and evidence for evolution.

- Qualitatively speaking, evolution is the slow change in the number of surviving genotypes from one generation to the next, and is reflected by a change in the phenotypes. Mutations introduce new genotypes that get tested over very long times, leading to slowly changing populations.
- Fossils remains of animals are always found in the same order in the geological layers. The oldest organisms are very simple and are found in the deepest layers. The fossil record is never found intermixed: human remains are never found with dinosaur remains, for example.
- Radioisotope dating involves measuring the radiation that is coming from a sample. The lower the radiation count is, the older the sample is. We can use radioisotope dating to confirm the age of fossil remains. The experiments confirm that organisms found in two different geological layers lived at different times – *in the order in which the layers were laid down.*
- Fossils and radioisotope dating has shown that not all organisms came into existence at the same time.
- Similarities in the fossil record also indicate a gradual change from one generation to the next, and by connecting the dots, we can get an excellent idea of how all species have evolved from primitive creatures.
- Cellular and biochemical methods have been used to measure the similarities in cellular function and DNA of today's organisms. These tests confirm the degree of similarities suggested by the fossil record. They show that, for example, chimpanzees and humans are very similar

$$p^2 + 2pq + q^2 = 1$$

Hidden in this is our Punnett square: the p^2 (which is the same as pp) is the probability of the top-left box. The q^2 is the probability of the bottom-right box, and the pq's are the two other boxes in the Punnett square. So basically equation 2 just says that the probability of the individual boxes of the Punnett square adds up to 100%. This, of course, is something we've encountered numerous times before in the heredity units.

> $p + q = 1$ tells us about the current population.
> $p^2 + 2pq + q^2 = 1$ tells us what happens when the population reproduces.

Next, let's step through each condition of the Hardy-Weinberg Principle. Here is the big test: **If all of these conditions apply to a population, then that population will not evolve**. If even one condition fails, evolution – the change in allele frequencies – will occur.

A) No mutations. This one is probably obvious. If mutations are occurring then it means a new allele will appear in the population, say allele r. If allele r is present then the combined frequencies of p and q will drop. For instance, frequencies of p and q of 70% and 30% might drop to, say, 69.99% and 29.99%, with the frequency of the new allele, r, being 0.02%. Since the frequencies changed, evolution has occurred. Therefore, *no mutation can ever take place or evolution must occur*. Unfortunately, mutations are impossible to stop.[17] As we saw in previous chapters, mutations occur at a rate of about one in one million base pairs. Some species have a somewhat lower rate.

---

[17] The impossibility of stopping mutations has a lot to do with the Second Law of Thermodynamics (the frequently quoted "order to disorder" rule of science). Basically, any sequence of DNA will undergo change because of this law. This is ironic, since mutations drive the wondrous variety of life. In a sense, disorder (i.e., chaos) is pushing evolution. Disorder has shaped our race into what it is, and we are kindly returning the favor to Nature.

Some have a higher rate. Unless the mutation rate drops to zero (again impossible) evolution must occur.

Given this impossibility, the rest of the conditions in the Hardy-Weinberg equilibrium are just overkill, but for the sake of thoroughness, let's examine the rest.

B) No Natural Selection. This statement means that Nature has to be fair, killing everything with equal probability. We know from common experience that Nature is not fair. For example, the "runts" of each litter are kicked out of the nest in many species. The slowest individuals are the first to get run down by the lion. Freak accidents kill perfectly healthy specimens. That's life. At any rate, since Nature is not fair, it means that some organisms are enjoying Nature's favor. That influences which genes make it to the next generation, which of course changes the allele frequencies and brings about evolution.

C) Infinitely large population. This one is not very obvious, but what it basically says is that small populations suffer from random genetic fluctuations (remember genetic drift?). Suppose that we have a small population of, say, four individuals. Further, say these individuals have p and q alleles at 50% each. Dumb luck alone might cause all the children of the four individuals to inherit all p alleles. The result is that the q alleles basically vanish from the population due to freak randomness. If the population is larger, like say 100 individuals, then the statistical probability of such extreme random fluctuations are much smaller. The larger the population is, the smaller the fluctuations. The only way to remove the random fluctuations entirely is by having an infinitely large population. The only way to have an infinitely large population is to have an infinitely large planet, which of course is impossible. Therefore, genetic drift alone will lead to evolution.

D) Everyone must breed. This condition says that everyone needs to participate in the breeding process and that no one can have reproductive problems. If only a fraction of the individuals breed, for

G

H

I

L

M

N

P

R

S

V

W

X

Y

Made in the USA
Monee, IL
04 November 2026

67377532R00066